ENGINEERING DISASTERS

CHERNOBYL

BY TODD KORTEMEIER

CONTENT CONSULTANT
Kate Brown, PhD
Professor of Science, Technology, and Society
Massachusetts Institute of Technology

Cover image: The Chernobyl Exclusion Zone is regularly tested for radiation levels.

Core Library

An Imprint of Abdo Publishing
abdobooks.com

abdobooks.com

Printed in China.
102019
012020

THIS BOOK CONTAINS RECYCLED MATERIALS

Cover Photo: Mondadori Portfolio/Getty Images
Interior Photos: Mondadori Portfolio/Getty Images, 1; V. Lysenko/TASS/Getty Images, 4–5; Shutterstock Images, 9; Shone/Gamma-Rapho/Getty Images, 12–13, 43; Barbara Bednarz/Shutterstock Images, 14; Keystone-France/Gamma-Keystone/Getty Images, 18; Rudi Blaha/AP Images, 20–21; Igor Kostin/Sygma/Getty Images, 24; Kirill Sergeev/Shutterstock Images, 26–27; LLNL/Science Source, 29; Patrick Landmann/Science Source, 31, 34–35; Emily Marie Wilson/Shutterstock Images, 39, 45

Editor: Marie Pearson
Series Designer: Ryan Gale

Library of Congress Control Number: 2019942092

Publisher's Cataloging-in-Publication Data

Names: Kortemeier, Todd, author.
Title: Chernobyl / by Todd Kortemeier
Description: Minneapolis, Minnesota : Abdo Publishing, 2020 | Series: Engineering disasters | Includes online resources and index.
Identifiers: ISBN 9781532190711 (lib. bdg.) | ISBN 9781532176562 (ebook)
Subjects: LCSH: Chernobyl Nuclear Accident, Chornobyl', Ukraine, 1986--Juvenile literature. | Disasters--Juvenile literature. | Engineering--Juvenile literature. | Nuclear power plants--Accidents--Juvenile literature. | Radioactive pollution --Juvenile literature.
Classification: DDC 363.179909--dc23

CONTENTS

CHAPTER ONE

A TEST GONE WRONG

It was a warm evening on April 25, 1986, in Pripyat, Ukraine, in the Soviet Union. Most of Pripyat's citizens worked at the nearby Chernobyl Nuclear Power Plant. Many residents were looking forward to May Day on May 1. A new amusement park was opening.

Chernobyl had four reactors. Reactors contain an element called uranium. Like all elements, uranium is made up of atoms. When uranium atoms split, the process creates a lot of heat. Water runs through the reactor, and the heat turns it to steam. The moving steam turns the fan blades on a turbine. As the turbine spins, it creates electricity.

Many people worked operating Chernobyl's reactors.

THE USSR

Chernobyl is located in Ukraine. Today Ukraine is its own country. But in 1986 it was a republic within the Union of Soviet Socialist Republics (USSR). Commonly known as the Soviet Union, the USSR existed from 1917 to 1991. After the USSR broke up in 1991, most of the republics became individual countries. Russia is the largest.

Chernobyl provided 10 percent of Ukraine's power.

In the early morning hours of April 26, workers at the power plant were preparing for a safety test. The test would happen in Reactor 4. The test was supposed to happen during the day on April 25. It had to get pushed back until the night shift. Many workers showed up not knowing they would be doing the test. Working the controls that night were shift foreman Alexander Akimov and engineer Leonid Tuptunov. Deputy chief engineer Anatoly Dyatlov was in charge.

The test was to see what would happen if the water supply to the reactor was cut off. The plant workers wanted to know if the turbine could still provide power as it slowed down. Normally that situation would set off emergency alarms for safety. These had to be turned off for the test.

The reactor's power also had to be reduced. It normally produced 3,200 megawatts of electricity. For the test, it had to be in the 700 to 1,000 megawatt range. But during the test, the power rate fell too low. At just after midnight, it was producing only 30 megawatts. Workers tried to boost the power level by pulling out control rods. Control rods control how fast uranium splits in a nuclear reactor. These rods absorb neutrons. Neutrons are produced when uranium splits. Putting in more control rods absorbs more neutrons. This slows down a reaction. Pulling them out speeds up the reaction.

Workers pulled out almost all the control rods. The reactor's instructions called for at least 15 control rods to stay in. But there was no explanation why. If workers couldn't control the nuclear reaction, the reaction could not be stopped. The fuel would get so hot that it would melt. This is called a meltdown.

Workers could not get the reactor back up to 700 megawatts. Akimov warned Dyatlov that it was unsafe to run the test with so little power. But Dyatlov ordered the crew to proceed.

RUNNING THE TEST

The test began at 1:23 a.m. The water pumps were shut off. Suddenly, the reactor's power began to spike. In seconds, it shot up to 530 megawatts. Temperatures in the core began to rise. A worker from the reactor building ran into the control room. The control rods, he said, were jumping out of the reactor floor. Heat and pressure were building in the reactor.

Akimov pressed the emergency button. All control rods began to move back into the reactor. That should have stopped the test and shut down the reactor. But instead, power spiked up to 120 times its full power.

A NUCLEAR REACTOR

A nuclear reaction generates heat, which makes water turn to steam. This steam powers a turbine, which generates electricity. Control rods in the reactor control how much heat the core is producing. How does this help you understand how a nuclear reactor works?

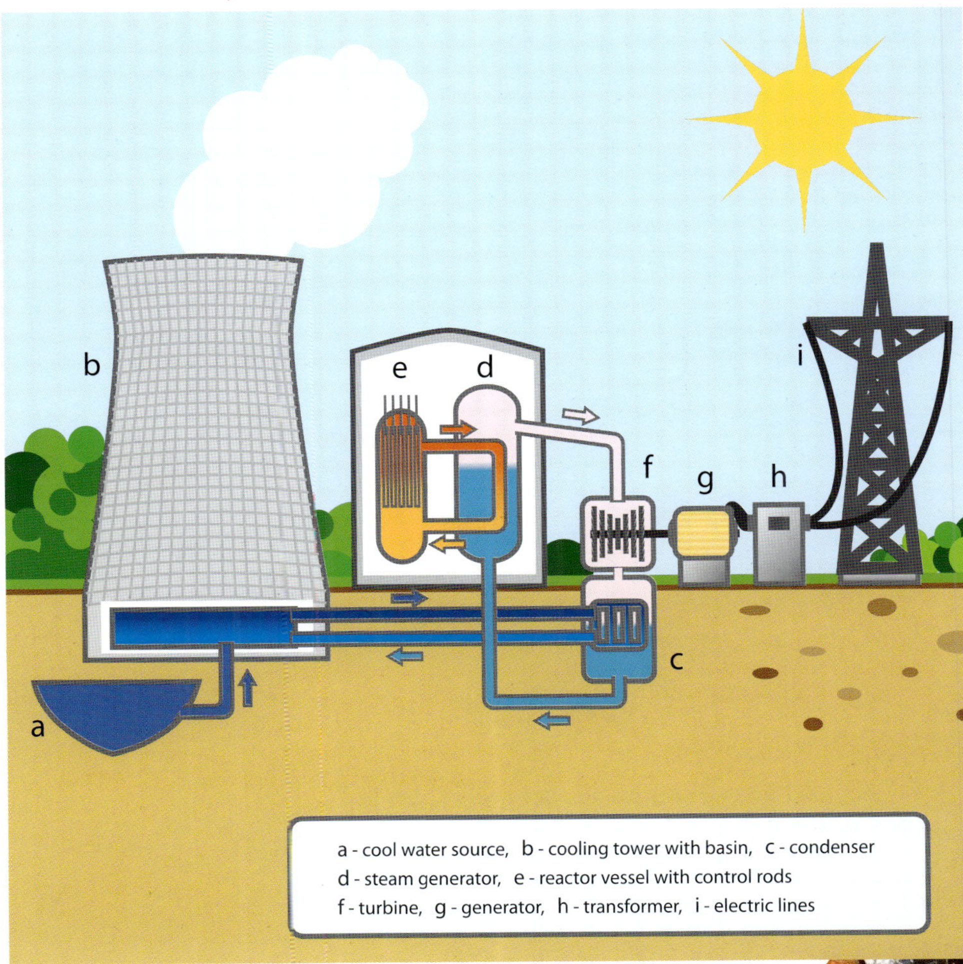

The intense heat made a lot of steam. The pressure from that steam blew the 1,000-ton (900-metric-ton) cover off of the reactor and through the roof above.

A fire started in the destroyed Reactor 4 building. Radioactive material was exposed to the air. It glowed crimson in the night sky, visible to the residents of Pripyat. It was symbol of a dangerous energy that would kill anyone unfortunate enough to come near it.

PERSPECTIVES

ALEXANDER YUVCHENKO

Nuclear power was a point of pride for Soviet citizens. Working at a plant like Chernobyl was a prestigious job. Alexander Yuvchenko was one such employee. "[Chernobyl] was one of the best stations in the Soviet Union, [Pripyat] was a good town to live in," Yuvchenko said. "And it was a good wage. Being a nuclear engineer was a prestigious career—in those days." Yuvchenko was working the night of the disaster. "When I heard the thud I thought it was something very heavy that had fallen. After that I didn't know. I thought that maybe war had begun."

STRAIGHT TO THE SOURCE

Nadezhda Petrovna Vygovskaya was a resident of Pripyat. She describes the night of the disaster:

> *I can still see the bright-crimson glow, it was like the reactor was glowing. This wasn't any ordinary fire, it was some sort of shining. It was pretty. I'd never seen anything like it in the movies. That evening everyone spilled out onto their balconies, and those who didn't have them went to friends' houses. We were on the ninth floor, we had a great view. People brought their kids out, picked them up, said, "Look! Remember!" . . . We didn't know that death could be so beautiful. Though I wouldn't say that it had no smell—it wasn't a spring or an autumn smell, but something else, and it wasn't the smell of earth. My throat tickled, and tears came to my eyes.*

Source: Melissa Block and Svetlana Alexievich. "'Voices from Chernobyl': Survivors' Stories." *NPR*. NPR, April 21, 2006. Web. Accessed July 26, 2019.

What's the Big Idea?

Read the passage carefully. What is the main idea? Name two or three details that support the main idea.

CHAPTER TWO

A LACK OF RESPONSE

At 1:32 a.m., a call came in to the Pripyat fire station. There was a fire up the road at Chernobyl. Thinking it was a regular fire, the firefighters loaded up their equipment and headed to the plant.

They soon realized it was no ordinary fire. Chunks of graphite were scattered on the ground. The only way that could happen was if they had been ejected from the core. The air above the plant was glowing red. And the firefighters had the strange taste of metal in their mouths. They were receiving very high doses of radiation.

An aerial photo shows Reactor 4 a few days after the meltdown.

A monument in Pripyat was dedicated to the firefighters of Chernobyl.

But the fire had to be put out. All the men had were their water hoses. That could do nothing for the burning core. But by 3:00 a.m., they had contained the fire to Reactor 4. Within a half an hour, some of the men started throwing up. They were taken to the city's clinic. Within a week, six of the 28 firefighters died of acute radiation poisoning. Those who survived were severely injured by radiation. They were kept away from their families for months. Twenty of the firefighters eventually died from radiation.

KEEPING A SECRET

Mikhail Gorbachev received an early morning phone call on April 26. Gorbachev was the General Secretary of the Soviet Union. He was the most powerful man in the country. The phone awakened him at 5:00 a.m. There had been an accident at Chernobyl, the caller said. But the accident was

PERSPECTIVES

LEONID TELYATNIKOV

Leonid Telyatnikov was a lieutenant colonel in the Pripyat fire department. He was one of the men sent to Chernobyl after the disaster. It was supposed to be his day off. But Telyatnikov did his duty along with his fellow firefighters. Telyatnikov died in 2004 at the age of 53 from cancer. It was likely due to the radiation dose he got at Chernobyl. "Chernobyl showed us that some people are not as conscientious and upright as they should be," he said. "Otherwise the accident would not have happened."

minor. And the core was intact. However, no one fully understood how bad things were yet.

In Pripyat, residents could not ignore the warning signs. Police and the military were in the streets. Some were wearing gas masks. Trucks were spraying down the roads with green chemical liquid. Planes and helicopters flew overhead. Schools were closed and residents were told to stay inside.

Gorbachev ordered an investigation. On the night of April 26, government officials met. They knew that Chernobyl was in full meltdown. They discussed what to do. They knew they had to keep it a secret. But radiation levels in Pripyat continued to rise. Scientists on the investigation convinced government officials to start an evacuation.

THE EVACUATION BEGINS

Quietly, the government started issuing evacuation orders on April 27. First to be evacuated were the 50,000 residents of Pripyat. Some of them had only

50 minutes to prepare. The order did not seem very serious. People were told the evacuation was temporary. They were told to only bring basic items. But after they left, the area around Chernobyl was closed off. Only those with passes were allowed in.

THE COLD WAR

Part of the reason for Soviet secrecy was the ongoing Cold War. The Soviet Union and United States were the two most powerful countries after World War II (1939–1945). They became rivals. They never declared war. But they constantly sought to beat one another in various ways. The race to outer space was one example. Admitting to the Chernobyl disaster was very embarrassing for the USSR. The Soviets felt it showed their nuclear power was inferior to that of the Americans.

The evacuation began at 2:00 p.m. Buses came and picked up all citizens of Pripyat. By 4:30 p.m., they were almost all gone. Just under 5,000 people stayed behind to run the still-working parts of the power plant. But otherwise, Pripyat became a ghost town.

As residents were evacuated, they were checked for radioactivity.

A SECRET GETS OUT

Chernobyl is more than 620 miles (1,000 km) from the Forsmark Nuclear Power Plant in Sweden. But the radiation had already traveled that far by April 28. The Soviets could pretend the disaster hadn't happened. But they couldn't hide the evidence. Workers entered the Swedish plant Monday morning. They had been away from the station for the weekend. But when they were checked before starting the day, they set off radiation detectors. The radiation had to have come from somewhere else. Swedish authorities checked with other plants around Europe. These plants reported high radiation levels too. The reports formed a trail to the southeast. That trail ended at Chernobyl.

The Swedish authorities asked the Soviets, who denied any accident. So the authorities said they would make a report to the International Atomic Energy Agency. That made the Soviets nervous. It was only then that the Soviets admitted to the disaster. An announcement aired on the news the evening of April 28. The statement said the accident was under control. Help was on the way to victims. And the government was investigating. But things were hardly under control. The cleanup was just beginning. Thousands of people were risking their lives in the effort.

EXPLORE ONLINE

Chapter Two talks about firefighters and other workers who were exposed to radiation. The website below talks more about the effects of the radiation. What new information did you learn?

CHERNOBYL CHILDREN INTERNATIONAL: ABOUT CHERNOBYL

abdocorelibrary.com/chernobyl-disaster

CHAPTER
THREE

THE LIQUIDATORS

A Soviet commission investigated the disaster. Valery Legasov, a nuclear chemist, was on the commission. Many Soviet government officials did not understand what they were dealing with. Legasov did.

The Soviet response was chaotic. The investigation team helped make some order. After the fire was put out, something had to be done about the radiation. Officials sent thousands of people to handle the cleanup. Called the liquidators, they were firefighters, military members, miners, janitors, and more. Approximately 600,000 liquidators served in

Valery Legasov was the deputy director of an atomic energy institute in Moscow, Russia, at the time of the Chernobyl disaster.

some way. Their jobs varied widely. Some had to clean up radioactive debris. Others cleared radioactive forests and buildings.

PERSPECTIVES
ARKADY FILIN

One of the tasks for the liquidators was cleaning up the soil. The earth around Chernobyl was contaminated by radiation. The top layer had to be scooped up and then buried. Otherwise it would keep giving off radiation. Arkady Filin was one of the men assigned to this duty. He recalled a strange task. "We buried the forest," he said. "We sawed the trees into 1.5-meter [5-foot] pieces and packed them in cellophane and threw them into graves. . . . We buried the earth." But burying the topsoil only helped for a couple of weeks.

The investigation team ordered radiation-absorbing material to be dropped on the damaged reactor. In all, 5,000 tons (4,500 metric tons) of sand and other materials ended up being needed.

It turned out the explosion was only the beginning of the disaster. Underneath Chernobyl was a pool of water for cooling the reactor. The reactor continued to burn. It burned through the floor of the plant. It got closer to the water.

The intense heat of the nuclear fuel could cause a steam explosion if it hit the water. This explosion would be much worse than the original blast. The water had to be drained. Three volunteers risked their lives. They went under the plant and opened valves to drain the pool. Despite the danger, all three men survived.

RADIOACTIVE ROOF

Cleanup continued on a massive scale. There was a growing problem. Something needed to be done about the radioactive material that had landed on the roof of nearby Reactor 3. Officials had to contain it. They needed to move the debris from Reactor 3 to the ruins of Reactor 4. Then they would construct a building over Reactor 4. That would shield most of the radiation.

First the Soviets tried robots. But they all failed because of radiation. There was only one option left.

High radiation levels on the roof of Reactor 3 left white streaks on the film of the photographs taken of the liquidators.

People had to do the work. Five thousand people cleared the roof by hand. Their exposure was limited. They worked in shifts of less than 90 seconds. A team of workers would run out, toss a shovelful of debris off the roof to Reactor 4, and then run back. Each worker only did this once. But each worker received at least five times a safe yearly dose of radiation in that small time.

THE SAVIORS OF CHERNOBYL

The roof workers were just one part of the liquidator force. All liquidators were exposed to high levels of radiation. Thousands of them later died of cancer. Many thousands more were disabled by radiation-caused diseases. Without the liquidators, the disaster would certainly have been worse.

LIFE AFTER LIQUIDATION

The liquidators were hailed as heroes in the Soviet Union. But their compensation did not equal their sacrifice. They received payments from the government. But after the Soviet Union collapsed in 1991, the money lost its value. All received military medals for their service. A monument was also built to honor the liquidators in Chernobyl. It was dedicated "to those that saved the world."

CHAPTER FOUR

DEMANDING ANSWERS

The exclusion zone around Chernobyl eventually grew to 19 miles (31 km) wide. Approximately 115,000 people were evacuated in 1986. Another 220,000 were resettled in the years that followed. But many had already been exposed to dangerous amounts of radiation.

In early May, workers began to pour concrete to build a containment system around Chernobyl. On May 14, Gorbachev spoke publicly about the disaster for the first time. He said there were nine deaths and nearly

The original structure built around Reactor 4 was called the sarcophagus.

300 people in the hospital. But he claimed the worst was over. International reports, he said, were lying about how bad it was.

THE SARCOPHAGUS

Radiation continued to pour out of Chernobyl in the months that followed the accident. The best way to stop it was by containing the exposed reactor. It took workers just 206 days to build a concrete and steel structure over the damaged reactor. Workers could only work for a few minutes at a time. The structure was called a sarcophagus. A sarcophagus is another word for a coffin.

The international nuclear community wanted answers. Why did this happen? What could be done to make sure it never happens again?

PART OF THE ANSWER

The Soviets blamed human error for the disaster. They said the plant workers who carried out the test made mistakes and ignored safety precautions. A trial was held in July 1987. Plant director Viktor Bryukhanov,

RADIATION AROUND THE GLOBE

Both human error and faulty engineering led to the Chernobyl disaster. The Atmospheric Release Advisory Capability simulated how far the radiation (pink) had dispersed just ten days after the meltdown. How does this map help you understand the scope of the disaster? How does it demonstrate the importance of careful engineering for nuclear reactors?

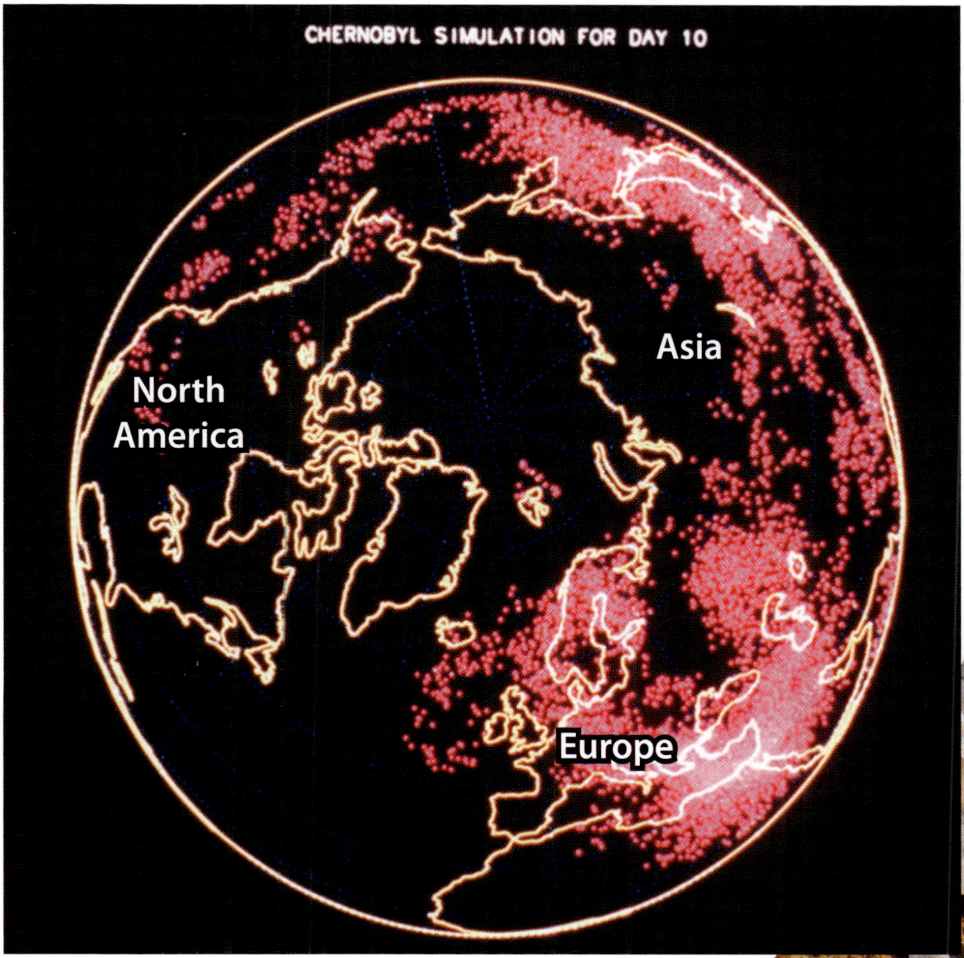

PERSPECTIVES

VALERY LEGASOV

Valery Legasov worked to find out what happened at Chernobyl and how to keep it from happening again. Legasov became depressed after Chernobyl. He was disappointed that the Soviet Union kept hiding things. His fellow scientists did not respect his work. He struggled with the after-effects of radiation. He died by suicide in 1988. Before his death, he wrote, "When one considers the chain of events leading up to the Chernobyl accident . . . it is impossible to find a single culprit, a single initiator of events, because it was like a closed circle."

chief engineer Nikolai Fomin, and deputy chief engineer Anatoly Dyatlov were sentenced to ten years of hard labor, though they didn't serve all of that.

Human error was certainly a factor. The operators that night broke many rules. They operated the reactor in an unsafe way. But it was due to more than just a few people. Safety procedures were often ignored at

The control room for Reactor 4 was abandoned after the meltdown.

Chernobyl and other plants in the Soviet Union. People took risks. Management at the plant wanted the test completed. The workers were committed to carrying it out, even in an unsafe way. Any workers who protested were overruled. But there was a lot the workers did not know about the reactor's design flaws too.

DESIGN FLAWS

There were many decisions that could have prevented Chernobyl. But because of the reactor's design, some factors were out of the workers' hands. Chernobyl was a type of RBMK reactor. RBMK reactors were unstable. But they were cheaper than other types, and they produced a lot of power. Other reactors slow down when there is less water flowing through the core. RBMK reactors produced more power. This creates more heat, which produces more steam. If unchecked, this reaction keeps speeding up.

In addition, when Leonid Tuptunov pressed the shutdown button, it actually increased the power.

The control rods were designed to absorb neutrons. But the tips were covered in graphite, which actually speeds up the reaction. Tuptunov and the workers did not know this. This led to the meltdown.

These problems were not new. Similar accidents had happened before in the Soviet Union. There had been many at Chernobyl before. Some had released radiation. Nobody in power made any changes. And these safety concerns were not communicated to the workers at the plant. It was a recipe for disaster.

FURTHER EVIDENCE

Chapter Four describes the causes of the Chernobyl disaster. What was one of the main causes? What evidence is there for that cause? Read the facts at the article below. Does the information from this website support the evidence in the book? Does it introduce new evidence?

BACKGROUNDER ON CHERNOBYL NUCLEAR POWER PLANT ACCIDENT

abdocorelibrary.com/chernobyl-disaster

CHAPTER FIVE

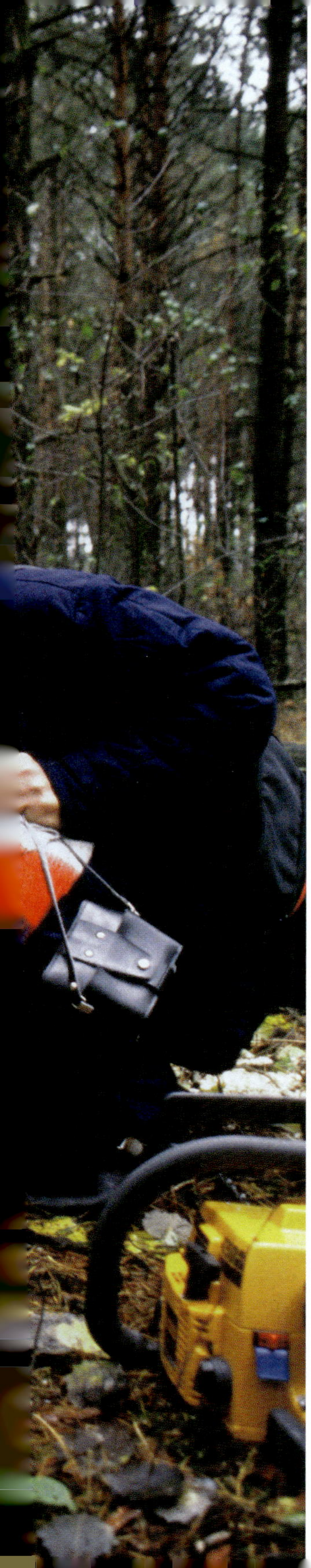

AFTERMATH

The Soviets could not ignore RBMK design flaws anymore. Safety standards were revised. Changes were made to the existing reactors. Material was added to the cores to reduce reactivity. This would help prevent a build-up of power with a lack of water.

Many safety measures had been turned off during the fateful Chernobyl test. Systems were installed to prevent turning these systems off while a reactor was running. And the emergency shutoff system was improved so it would kick in faster. The control rods were fixed so that they could not speed up

Scientists are still studying the effects of the Chernobyl disaster. One way they do this is by measuring radiation in the nearby forest.

the reaction. As of 2018, Russia still used ten RBMK reactors. They provided 31 percent of the country's nuclear electricity.

FUKUSHIMA

Nuclear disasters are rated based on how serious they are. The scale goes up to seven. Only two disasters have rated a seven: Chernobyl and the Fukushima disaster in 2011. The Fukushima-Daichi nuclear power plant is on the east coast of Japan. It was heavily damaged by an earthquake and tsunami. All three of its reactors melted down. It took Japan two months to admit to the meltdowns. But much more radiation was released at Chernobyl. Fukushima was also a serious accident, but no direct deaths were reported. The long-term effects remain to be seen, however.

QUESTIONING NUCLEAR SAFETY

Issues in reactor designs could be fixed. But nuclear power still has dangers. The accident raised concerns about nuclear power worldwide. Was it worth the risk?

Chernobyl was not the first major nuclear disaster. In 1979, there was a partial meltdown in the United States.

Accidents at nuclear bomb facilities leaked radioactive material. After the far worse accident at Chernobyl, several countries stopped building nuclear power plants. Others changed their safety requirements. These rules made it hard to build new power plants quickly. Once seen as the future of energy production, nuclear power was never the same after Chernobyl.

CHERNOBYL TODAY

At Chernobyl, the plant's three healthy reactors continued to provide power. By 1988, the plant was back to producing its normal power levels. Then the other three reactors were shut down one by one. The plant was fully closed in 2000. The fuel still has to be carefully removed from the site. Chernobyl is predicted to be fully dismantled by 2064.

But parts of the Chernobyl Exclusion Zone still won't be safe by then. Some will still be radioactive for hundreds of years. Reactor 4 won't be safe for 20,000 years.

As for humans, Pripyat remains a ghost town. The area is safe enough for people to visit briefly, though. Several films and television shows have been made about the area.

PERSPECTIVES

HANNA ZAVOROTNYA

Most people evacuated from the Chernobyl area never returned. But a few did go back. Some of them still live in the exclusion zone. They are known as *samosely*, meaning "self-settlers." Living in the exclusion zone is illegal. But Ukraine's government supports these people with basic supplies. Fewer than 150 samosely are left, most of whom are women. Hanna Zavorotnya was interviewed in the 2015 documentary *The Babushkas of Chernobyl*. She remembered being asked to leave her homeland. "Shoot us and dig the grave," she told a Soviet soldier. "Otherwise, we're staying."

THE HUMAN COST

The true death toll of Chernobyl is almost impossible to know. The official Soviet total is 31 people who died as a direct result of the explosion. The United Nations puts the number at 54.

In 2017, the Chernobyl New Safe Confinement structure was built over the crumbling sarcophagus. A monument honored the builders of the sarcophagus.

But radiation does not always kill people right away. It can cause other diseases such as cancer later on. In 2006, the World Health Organization estimated 9,000 more people would get cancer as a result of Chernobyl. But even that report admitted the number was impossible to predict. Some estimates have soared even higher. A scientific survey in 2011 suggested 27,000 to 108,000.

What is not debatable is the massive disruption to human life. Hundreds of thousands of people were forced to move. Several hundred thousand more were asked to clean up the mess. Millions around Europe were exposed to radiation.

All of it was preventable. Chernobyl happened because the reactor didn't have necessary safety features and people took risks. Reactors can be fixed. Safety manuals can be updated. But it requires a dedicated effort to prevent disaster in the first place.

STRAIGHT TO THE
SOURCE

Mark Z. Jacobson, director of Stanford University's Atmosphere/Energy Program, wrote about the risks of nuclear energy:

> *To date, 1.5% of all nuclear power plants ever built have melted down to some degree. . . . The nuclear industry has proposed new reactor designs that they suggest are safer. However, these designs are generally untested, and there is no guarantee that the reactors will be designed, built and operated correctly or that a natural disaster or act of terrorism, such as an airplane flown into a reactor, will not cause the reactor to fail, resulting in a major disaster.*

Source: Mark Z. Jacobson. "The 7 Reasons Why Nuclear Energy Is Not the Answer to Solve Climate Change." *Leonardo DiCaprio Foundation*. Leonardo DiCaprio Foundation, June 20, 2019. Web. Accessed September 17, 2019.

Changing Minds

Take a position on whether you think people should continue using nuclear power. Then imagine that your friend has the opposite opinion. Write a short essay trying to change your friend's mind. Include facts and details that support your position.

FAST FACTS

- Chernobyl was a nuclear power plant located near Pripyat in modern-day Ukraine.
- Pripyat was a city primarily filled with residents who worked at the plant and the families of those workers.
- In the early morning of April 26, 1986, workers at the plant were conducting a safety test. The test went wrong, leading to an explosion in one of the nuclear reactors that released radioactive material into the air.
- The Soviet Union government at first refused to acknowledge the incident, even keeping it a secret from residents of Pripyat. These citizens were not evacuated for more than a day, at which time an exclusion zone was set up around the plant.
- Cleanup efforts were carried out by 600,000 Soviet citizens called liquidators. These liquidators performed very dangerous work cleaning up the contaminated areas.
- Thousands of people have since died of health issues caused by radiation.
- While the Soviet Union first blamed the plant workers for the accident, it was also proven the reactor had fatal design flaws that the workers didn't know about.
- Other reactors were fixed, but nuclear power remains a controversial source of energy.

STOP AND THINK

Tell the Tale

Chapter Five mentions that tourists now visit the Chernobyl Exclusion Zone. Imagine you are taking a trip to this place. What kind of things might you see? How would you stay safe?

Surprise Me

Chapter One discusses how nuclear reactors work. After reading this book, what two or three facts about nuclear power did you find most surprising? Write a few sentences about each fact. Why did you find each one surprising?

Say What?

Studying nuclear reactors can mean learning a lot of new vocabulary. Find five words in this book you've never heard before. Use a dictionary to find out what they mean. Then write the meanings in your own words, and use each word in a new sentence.

Dig Deeper

After reading this book, what questions do you still have about the Chernobyl disaster? With an adult's help, find a few reliable sources that can help you answer your questions. Write a paragraph about what you learned.

GLOSSARY

atom
the smallest part of any element

element
the most basic part of any substance, an element contains only one type of atom

evacuation
going away from a particular area

exclusion zone
an area where human activity is restricted

foreman
a person who is in charge

megawatt
a unit of electricity; one megawatt can power approximately 650 homes

meltdown
when a nuclear core overheats and begins to melt

neutron
a particle that is part of an atom and has no positive or negative charge

radiation
energy that is given off by particles

radioactive
gives off radiation

shift
a period during which a person is working

turbine
a machine that spins to generate electricity

ONLINE RESOURCES

To learn more about Chernobyl, visit our free resource websites below.

Core Library CONNECTION
FREE! COMMON CORE MULTIMEDIA RESOURCES

Visit **abdocorelibrary.com** or scan this QR code for free Common Core resources for teachers and students, including vetted activities, multimedia, and booklinks, for deeper subject comprehension.

Booklinks NONFICTION NETWORK
FREE! ONLINE NONFICTION RESOURCES

Visit **abdobooklinks.com** or scan this QR code for free additional online weblinks for further learning. These links are routinely monitored and updated to provide the most current information available.

LEARN MORE

Marquardt, Meg. *Nuclear Energy*. Minneapolis, MN: Abdo Publishing, 2017.

Rissman, Rebecca. *The Chernobyl Disaster*. Minneapolis, MN: Abdo Publishing, 2014.

INDEX

About the Author

Todd Kortemeier is a children's book author and sportswriter from Minneapolis. He and his wife live with their dog west of the city.